◄I'M GOING TO BE A►
Farmer

◄I'M GOING TO BE A►
Farmer

by Edith Kunhardt

The author thanks Harry, Barbara, Nathan, and Meredith Ludlow of Bridgehampton, New York, for inspiring this book, and for their warm hospitality and friendship. Many thanks also to Gurden and Marjorie Ludlow.

ISBN 0-590-25482-0

Text and photographs copyright © 1989 by Edith Kunhardt.
All rights reserved. Published by Scholastic Inc.
CARTWHEEL BOOKS and the CARTWHEEL BOOKS logo are registered trademarks of Scholastic Inc.

12 11 10 9 8 7 6 5 4 3 2 1   6 7 8 9/9 0 1/0

Printed in the U.S.A.   24

First Scholastic printing, May 1996

Cartwheel
·B·O·O·K·S·®
SCHOLASTIC INC.
New York   Toronto   London   Auckland   Sydney

My name is Nathan. This is my house.

My mom and dad are farmers. They grow their own food. And they raise potatoes to sell at the market.

I want to be a farmer, too.

Every morning we get ready to go out. I put on my boots. My little sister, Meredith, puts on her sneakers. Daddy and Mommy get ready, too. Grandpa gets his pail.

We go to work right away.

Oh! Oh! What do I see?

Good morning, baby chicks!
Did you keep warm under the
brooder? Peep, peep, baby chicks.
I see you.

Mommy pitches the old straw out of the barn. She puts in nice, clean straw. She gives the sheep hay to eat.

The pigs chew corncobs from our dinner last night. I rake the floor.

Gobble, gobble, gobble! The turkeys all say
"gobble" at once. They shake their red wattles.
Meredith and I go to see the goose and the
gander. The gander sticks out his neck and
hisses at me. Go away, gander. He goes away.

There is Red, a horse.
It tickles when he nibbles
the grass out of my hand.
Meredith pats Chester.

I help out a lot. I fix things. And I ride in the tractor with Daddy. It's air-conditioned in the cab.

Potatoes grow under the ground. This is what they look like when I pull them up.

The potato digger drives
along and digs up the potatoes.
The potatoes go into the truck.
Then they come out of the truck
onto the conveyor belt. Then
they go to the market.

Sometimes Meredith and I hoe the vegetable garden. I know how to pick a pepper right off the plant. This one isn't ready yet.

Oh, boy! Popcorn. We help
Mommy pick it off the cornstalks.
We fill up a whole basket.

Later we will dry the corn
and pop it. With lots of butter,
it tastes delicious!

We raise all kinds of things
on our farm. We grow tomatoes,
sunflowers, and pumpkins.

It's getting late now.
We hurry home.

Dinnertime! We have pork chops and broccoli and guess what—baked potatoes!

Grandma and Grandpa come over for dessert. After they go home, Daddy reads to us. We read a book about farms.

I'm going to be a farmer.